FLORA OF TROPICAL EAST AFRICA

CABOMBACEAE

B. Verdcourt

Aquatic herbs with perennial sympodial rhizomes; stems long and slender, coated with mucilage. Leaves alternate, the floating ones peltate, the submerged ones finely divided or absent. Flowers rather small, hermaphrodite, axillary, solitary and regular. Sepals 3,* petaloid. Petals 3, hypogynous. Stamens 3–18, with extrorse anthers dehiscing longitudinally. Carpels 2–18, completely free, with a simple narrow stigma but very reduced style; ovules 1–3, parietal, pendulous, anatropous*. Fruiting carpels indehiscent, 1–3-seeded. Seeds with fleshy endosperm and perisperm but no aril.

A family of two genera, one occurring in both the Old and New Worlds and the other occurring naturally only in America; it is sometimes regarded as a subfamily of Nymphaeaceae. *Cabomba aquatica* L. has been cultivated in Kenya as an aquarium plant (e.g. Nairobi, Oct. 1955, *Bally* 10428 !).

BRASENIA

Schreb. in L., Gen. Pl., ed. 8: 372 (1789)

Floating leaves peltate; submerged leaves absent. Sepals and petals persistent in fruit. Stamens 12–18. Carpels 6–18. Testa with cells of outer layer very characteristically elongated radially.

A genus with only a single widespread recent species.

B. schreberi *J. F. Gmel.* in L., Syst. Nat., ed. 13, 2: 853 (1791); Chifflot in Annls. Univ. Lyon, Fasc. 10: 39–49 (1902); Exell & Mendonça in C.F.A. 1: 45 (1937); Flora U.R.S.S. 7: 5, t. 1/3 (1937); Gleason, New Britton & Brown Illustr. Fl. 2: 148, fig. (1952); Wild in F.Z. 1: 173, t. 25 (1960); F.P.U.: 46 (1962); Meusel & Mühlberg in Hegi, Illustr. Fl. Mitteleuropa 3 (3), 1: 7, fig. 4 (1965) [map of recent and fossil distribution]. Type: United States of America, New Jersey, Hope (M, holo., *fide* Merxmüller in Wild, *loc. cit.*)

Apparently glabrous aquatic with all submerged parts covered with a profuse clear slimy jelly-like mucilage; the outsides of the sepals and petals, the undersurfaces of the leaves, submerged petioles, pedicels, etc., are covered with minute purplish glandular hairs which may secrete this mucilage. Leaf-blades round or broadly elliptic, 2·5–11·5 cm. long, 1·5–7 cm. wide, entire, green above, usually reddish-brown or purplish beneath with 12 nerves radiating from the centre; petiole-length depending on the depth of the water, 0·05–1 m. long or more, slender, ± 1·5 mm. in diameter. Pedicels reddish-brown, 4–5 cm. long, 2 mm. in diameter. Sepals purplish- or brownish-red, elliptic or narrowly oblong, 0·9–1·4 cm. long, 3–6 mm. wide, rounded at

---

* Chifflot mentions that there can rarely be 4 petals and sepals and up to 36 stamens. Most authors describe the ovules as orthotropous in Cabombaceae and Nymphaeaceae but Baillon, Chifflot and others refute this.

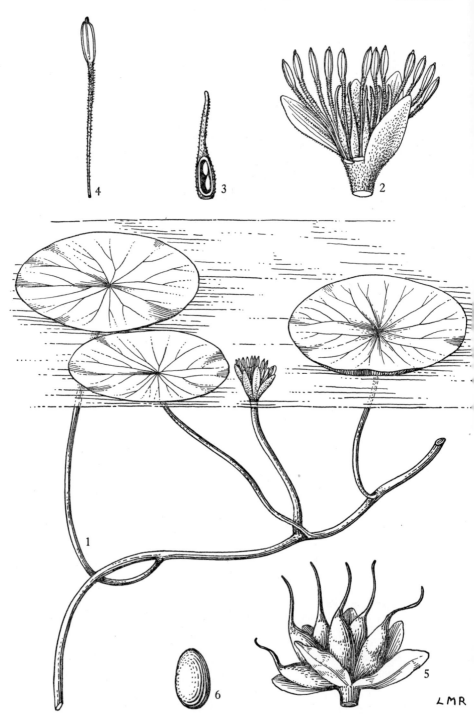

FIG. 1. *BRASENIA SCHREBERI*—**1**, habit, × ⅔; **2**, flower, with one sepal and one tepal removed, × 2; **3**, carpel opened to show ovules, × 3; **4**, stamen, × 3; **5**, fruit, × 2; **6**, seed, × 6. 1–4, from *Story* 4786; 5, 6, from *Gilges* 7. Reproduced with permission of the Editors of Flora Zambesiaca.

the apex. Petals crimson, purplish- or brownish-red, narrowly oblong or linear-oblong, 1·2–1·7 cm. long, 3–4 mm. wide, narrowly rounded at the apex. Filaments of the stamens filiform, 1 cm. long, papillose; anthers reddish, linear-oblong, 3·5 mm. long, apiculate. Carpels fusiform or narrowly ovoid, 5 mm. long, 1·5–2 mm. wide, sparsely papillose; stigma simple, 3–4·5 mm. long, papillose all over save for a narrow longitudinal area facing the perianth. Fruiting carpels ellipsoid or fusiform, 0·6–1 cm. long, 2–2·5 mm. wide; stigma persistent. Seeds pale brown, ellipsoid, 2·7–3·5 mm. long, 2–2·3 mm. wide.

UGANDA. Masaka District: E. side of Lake Nabugabo, 6 Oct. 1953, *Drummond & Hemsley* 4628!; Mengo District: Kyagwe [Kiagwe], Namanve Swamp, Sept. 1932, *Eggeling* 563! & Kampala, King's Lake, 21 July 1935, *Chandler & Hancock* 12!
KENYA. Ravine District: Equator, Lake Narasha, 16 Oct. 1953, *Drummond & Hemsley* 4802!
TANGANYIKA. Bukoba District: Kongolero Lake, Sept. 1931, *Haarer* 2184!; Njombe District: Ukinga, Bulongwa, 26 Jan. 1914, *Stolz* 2463!
DISTR. **U**4; **K**3; **T**1, 7; Zambia, Rhodesia, Angola, Botswana, South Africa (Transvaal), U.S.S.R. (Far East), India, Manchuria, Japan, Australia, N. America (U.S.A. and Canada), Mexico, Guatemala, British Honduras, Cuba, Guyana; also known from Tertiary and interglacial deposits in Europe
HAB. Lakes, swampy pools, etc.; 1120–2750 m.

SYN. *Hydropeltis purpurea* Michx., Fl. Bor.-Am. 1: 324, t. 29 (1803). Types: United States of America, S. Carolina and Tennessee, *Michaux* (P, syn.)
Brasenia peltata Pursh, Fl. Am. Sept. 2: 389 (1814); Oliv., in F.T.A. 1: 52 (1868), *nom. illegit.* Types: as for *Hydropeltis purpurea*
B. purpurea (Michx.) Casp. in Jorn. Sci. Acad. Lisb. 4: 312 (1873)*; R. E. Fries in Wiss. Ergebn. Schwed. Rhod.-Kongo-Exped. 1: 39 (1914)

* Usually seen as a reprint Jorn. Sci. Math. Phys. Nat. [Lisboa] No. 16: 1.

INDEX TO CABOMBACEAE